The Great
Testosterone Myth

Aniruddha Railkar

ISBN-9781534709409

Published by
10-10-10 Publishing
Markham, Ontario
CANADA

LEGAL DISCLAIMER
The advice in this book is for education and information purposes only. It is not intended to diagnose, treat, cure or prevent any disease. Before undertaking any exercise or other such physical activity, consult a physician first. The author and publisher of this book are not responsible for use, misuse or disuse of the information provided.

CONTENTS

This book is dedicated to my loving wife, Radha, and our adorable children, Revati and Shubham. Their love and support is the source of my inspiration.

ABOUT THE AUTHOR

Aniruddha Railkar was born and brought up in Mumbai, India. After getting his BSc degree in Chemistry from the University of Bombay, India, he came to the US. He has a MS in Pharmaceutical Sciences from West Virginia University in Morgantown, WV. He has a PhD in Pharmaceutical Sciences from the University of the Sciences in Philadelphia. Dr. Railkar has worked in the pharmaceutical industry for 18 years. He has written book chapters and been published in peer reviewed scientific journals. He has written 3 non-fiction books. He currently lives in Ambler, PA with his family.

FOREWORD

Information is a double-edged sword. Right and wrong information is everywhere, and especially with the growth of the Internet, delivery of information has exploded. However, it is up to you to interpret this information and apply it to your personal situation.

Dr. Aniruddha Railkar is a pharmaceutical scientist by trade and been in the pharmaceutical industry for eighteen years. *The Great Testosterone Myth* is a compilation of current information about low testosterone and the myth surrounding it. Dr. Railkar has performed a comprehensive literature search and presented the information in an unbiased and simple manner so that you can easily absorb and apply it.

This book is laid out in a way that is simple and easy to follow. It starts with a review of male anatomy and physiology and then describes the nature of the problem and how it is diagnosed. It also lists signs and symptoms and available treatments, both medical and non-medical. It is possible that you may be experiencing the effects of low testosterone. This book does a good job of explaining everything and discussing the pros and cons of treatment. Ultimately it is up to you to decide whether or not you want treatment. By reading this book, you are now armed with the knowledge to make an informed decision.

I highly recommend this book and look forward to other books from Dr. Railkar in the future.

Raymond Aaron
New York Times Best-selling Author

INTRODUCTION

So, if anatomy is destiny then testosterone is doom.

Al Goldstein

Testosterone is a steroid hormone found in human beings and vertebrate animals. It is secreted by the testicles in a male and to a small extent by the ovary in the female. Typically, the amount of testosterone in the adult male population is about 8 times higher than the corresponding amount in adult females. In males, testosterone is responsible for the development of reproductive organs. In addition, it is also responsible for the development of other sexual characteristics such as increase in muscle and bone mass and growth of body hair. Research has also shown that it prevents osteoporosis in males.

As we age, testosterone levels decrease naturally, about 1% per year after the age of 30 as shown in Figure 1. Decrease in testosterone causes decrease in muscle mass and libido.

However, if the testosterone level goes below 300 ng/dL (nanogram per deciliter) then the condition is classified as hypogonadism. About 30% of males suffer from this condition. Contrary to popular belief, men also have a biological clock. As soon as women reach menopause, the proverbial stroke of midnight, their estrogen levels drop drastically and the ovaries stop producing eggs. The decline in men is more slow and gradual. In women there are warning signals such as hot flashes, irregular periods, but in men the warning signs are not as pronounced. Some signs are slow metabolism, longer recovery periods from exercise such as weight lifting, lack of sexual desire and loss of libido. This condition is called andropause. At that time, a visit to a doctor followed by a blood test may indicate hypogonadism. Other factors such as genes, lifestyle (diet, exercise, stress, smoking, alcohol intake) obesity, diabetes and exposure to pollutants can also result in low testosterone.

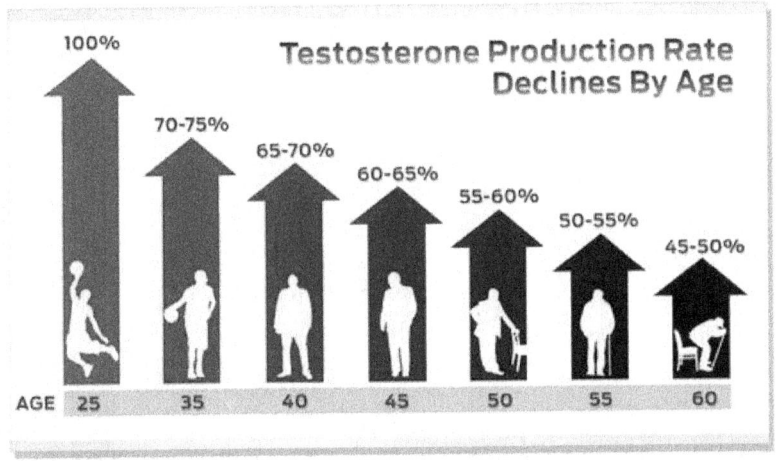

Figure 1: Decrease in testosterone as we age

It should be noted that testosterone is not the fountain of youth. Unfortunately, natural age related decrease in levels of testosterone has been exploited by certain segments of society. So called "wellness" clinics have sprung up all over the country to treat this condition. Direct to consumer advertising has added to the hype by making older males feel like they need to seek treatment to feel youthful again. In this book, an attempt has been made to look at all aspects of low testosterone and separate myth from fact. In addition a review of male anatomy and physiology, pharmaceutical treatments and their side effects, and natural remedies are discussed.

A REVIEW OF MALE ANATOMY AND PHYSIOLOGY

I am fascinated by the human body and all its evolutions

Jock Sturges

The human body is indeed a work of art. It is perfectly designed for what it is destined to do; nothing more, nothing less. Our body is made up of different systems that work in harmony with each other right from the day we were born. In order to understand the information presented in future chapters, a brief review of the male anatomy and physiology is necessary. However, the focus of this chapter will be the male reproductive system because it is predominantly affected by testosterone. The male reproductive system can be broadly divided into:

- External Organs
- Internal Organs

External Organs

The male reproductive anatomy is shown in Figure 2.

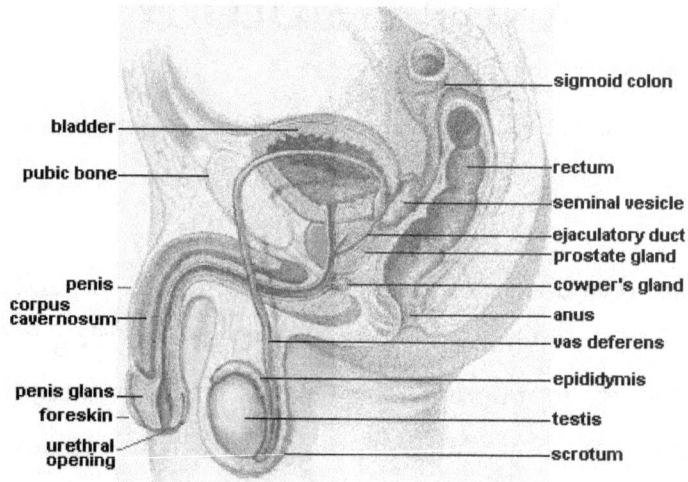

Figure 2: Male reproductive anatomy

The external organs are the penis, testes (or testicles) and the scrotum, the sack that holds the testicles.

Internal Organs

The internal organs are also shown in Figure 2. They are:

Epididymis- where sperm mature and are stored and then taken to the Vas Deferens

Vas Deferens- A 30 centimeter long tube, also known as sperm duct.

Accessory Glands- There are three accessory glands namely, seminal vesicles, prostate gland and bulbourethral or Cowper gland. Their function is to provide fluid and nourishment to the sperm.

WHAT IS TESTOSTERONE?

I think that testosterone is a rare poison.

Germaine Greer

Background

As mentioned earlier, testosterone is an anabolic steroid hormone. Anabolism is a process that increases protein production and helps build muscle and bone. Steroid is a chemical that has four rings of carbon and hydrogen along with other chemical groups. Its chemical structure is shown in Figure 3.

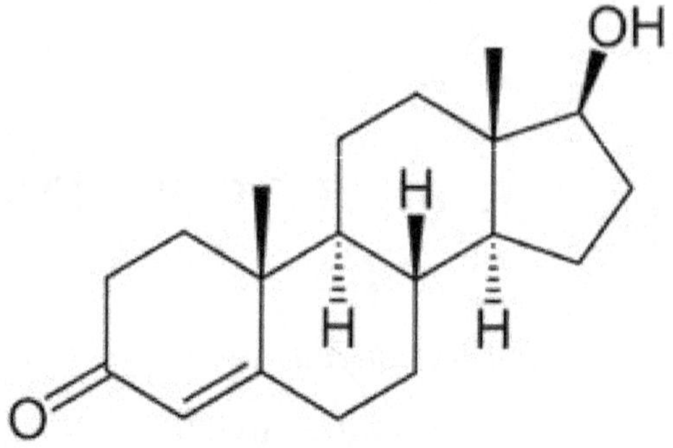

Figure 3: Chemical Structure of Testosterone

A hormone is a substance that is secreted in one part of the body, but acts in another part of the body. A hormone also affects more than one organ or organ system. Hormones are like our body's messengers. They communicate between different systems and make them work in harmony. Testosterone plays an important role in the development of male reproductive tissues. The secondary role is development of external sexual characteristics such as increased muscle and bone mass, growth of hair and general health. Osteoporosis which was believed to be a disease only affecting post-menopausal women, can also affect men who have low testosterone.

History

The history of testosterone can be traced as far back as the ancient Greek and Chinese civilizations. Aristotle is said to have written about the role of testosterone in reproductive biology. Scientific experiments were first carried out by John Hunter in London in 1786. But his experiments did not yield definitive results. Almost a century later, in 1849, experiments carried out by a German scientist called Arnold Adolph Berthold reignited the interest in testosterone. In 1889, Harvard professor, Charles-Edouard Brown-Séquard, injected himself with an extract from dog and guinea pig testicles. He reported that he felt more vigor and energy after the injection, but unfortunately the effect did not last along. There was also speculation that Brown-Séquard may have experienced a placebo effect. Due to the nature of his experiments and the possibility of placebo effect, Brown-Séquard was ridiculed by his peers and he abandoned all his efforts on testosterone. In 1927, a chemistry professor at the University of Chicago was able to extract testosterone from bull testicles. They found that when administered it remasculinized castrated roosters, pigs and rats. However, the extraction process was not very efficient. Forty pounds of bull testicles yielded 20 mg of testosterone. But in the 1930's three major European pharmaceutical companies started large scale programs on steroid research which laid the groundwork for synthesis of large quantities of testosterone. The name testosterone is derived from the words testicle, steroid and ketone. Another German scientist, Adolph Butenandt determined the structure of testosterone and two Swiss scientists, Leopold Ruzicka and A. Wettstein published the synthesis of testosterone. From 1930-1950 a lot of work was done on testosterone which indicated that it was a powerful chemical that plays an important role in muscle mass, strength and overall well-being.

Synthesis

Testosterone is produced in the Leydig cells in the testicles. The starting chemical for testosterone and other steroid hormones is cholesterol. The synthetic pathway for any hormone is complex. Many enzymes and other biochemicals are involved. However, testosterone synthesis consists of three major steps. Cholesterol is converted to Pregnenolone, which is subsequently converted to Androstenedione. Androstenedione is then converted to testosterone. The synthetic pathway can be summarized as shown below.

Cholesterol→Pregnenolone→Androstenedione
→Testosterone

Regulation

When levels of testosterone decrease, gonadotropin-releasing hormone (GRH) is released by the hypothalamus. GRH then stimulates the pituitary gland to release luteinizing hormone (LH) and follicle stimulating hormone (FSH). LH and FSH stimulate the testicles to produce testosterone. If testosterone levels are high, it provides a negative feedback to the hypothalamus and pituitary gland which inhibit the release of GRH, which in turn inhibits the release of LH and FSH as shown in Figure 4.

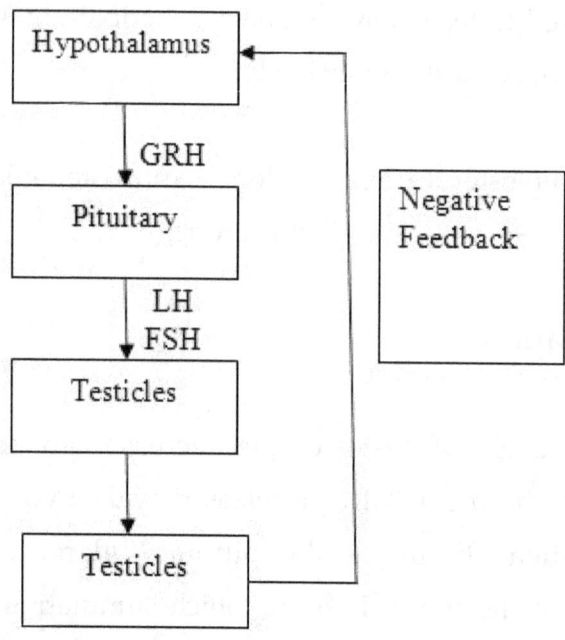

Figure 4: Synthesis of testosterone

Metabolism

About 90% of the testosterone is metabolized in the liver. A small percentage is metabolized by the prostate gland and skin. Fatty tissues convert about 0.2% of the testosterone to estradiol. The major metabolite of testosterone is dihydrotestosterone and it also has properties similar to testosterone. Typical male characteristics are seen because of testosterone and dihydrotestosterone together. However, their function is different. In puberty, testosterone induces the sex drive in men, enlargement of the penis, the production of sperm, increase of muscle mass and deepening of the voice. These effects are called anabolic effects.

Dihydrotestosterone is responsible for androgenic effects such as an increase of body hair, beard growth, acne, and at a later age for baldness and enlargement of the prostate. But going back to the metabolism, testosterone and dihydrotestosterone are both eventually converted to inactive chemicals and excreted in the urine and feces. Two other minor metabolites are etiocholanolone and androsterone. They are also eventually converted to inactive chemicals and excreted in the urine and feces. It can be depicted as:

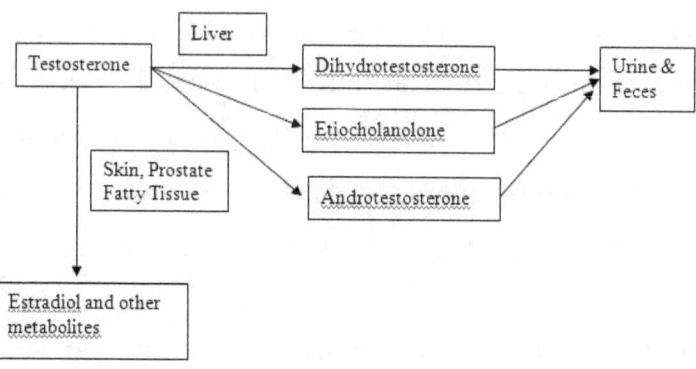

Figure 5: Metabolism of Testosterone

Effects

Testosterone plays a major role in the body predominantly in males, but also to a lesser extent in females. Effects of testosterone start as early as 4 to 6 weeks after conception. One is genital virilization or early development of genital characteristics and development of prostate and seminal vesicles. Testosterone levels in the second trimester are associated with gender formation, sometimes called feminization or masculinization of the fetus. Another effect of this is that the brain is "imprinted" to be male. In male infants, testosterone levels rise in the first few weeks of life, although the reason is unknown. At 4 to 6months of age, testosterone levels can be detected in male infants. Just before males reach puberty, as testosterone levels rise, they develop adult like body odor, skin and hair become oily and they can develop acne. There is appearance of pubic hair, hair on the upper lip, chin, and growth of sideburns. Also they experience a growth spurt and bone maturation. At puberty testosterone is responsible for enlargement of sweat glands, enlargement of the penis, more facial hair, chest hair, hair in the armpit and on legs. In addition, the voice becomes deeper, shoulders become broad and rib cage expands. Growth of the Adam's apple, growth of spermatogenic tissue in testicles, beginning of male fertility and definition of jaw bone, chin, nose and other facial bones also occur.

It has an effect on the endocrine, reproductive, cardiovascular and central nervous systems. In addition, it has an effect on other systems as well.

The Endocrine System

This system is made up of glands and it secretes hormones. Earlier, we discussed the positive and negative feedback loop (refer to Figure 4) involving the hypothalamus and the pituitary.

The Reproductive System

As early as seven weeks after conception, testosterone production can start and help in developing male reproductive organs. As the child grows, amount of testosterone produced increases significantly until it peaks in the mid-twenties and then starts to decline (see Figure 1). Then there are the anabolic effects of testosterone such as increase penis size and increase in muscle mass. The androgenic effects of dihydrotestosterone such as growth of body hair and development of acne were also discussed. Normal development of the prostate gland is dependent on testosterone. In order for testosterone to exert an effect, it needs to be converted to dihydrotestosterone.

Testosterone can also help with treating erectile dysfunction. Erectile dysfunction can be caused by emotional, vascular, neurological and pharmacological factors. The treatment of choice is administration of phosphodiesterase-5 (PDE-5) inhibitors. But for some people PDE-5 inhibitors don't work. Adding testosterone to PDE-5 inhibitors works.

Sexuality

As a result of increase in testosterone production, there is an increase in sexual desire. Testosterone works on the sexual centers in the brain. Sexual desire and activity can increase testosterone and a lack of sexual activity can reduce the levels of testosterone. Male fertility and libido are maintained by testosterone.

Central Nervous System

The hypothalamus and pituitary are in the brain. When testosterone levels go down, the hypothalamus signals the pituitary, which in turn signals the testicles to produce testosterone. Similarly when testosterone levels rise, because of the negative feedback, the hypothalamus tells the pituitary, which tells the testicles to stop production. There are testosterone receptors in the brain. This influences how we think and act. Hence, testosterone is responsible for effects such as anger, dominance, competitiveness, assertiveness, confidence and ability to take risk, although it is not the only determining factor. There is a modest relationship between levels of testosterone and aggression and criminality. There is a direct relationship between testosterone and dominance. Studies have shown that the most violent criminals tend to have high testosterone levels. In a study of financial traders and brokers, higher levels of testosterone were related to risky trades. Testosterone regulates cognitive and physical energy.

Circulatory System

The production of red blood cells in the bone marrow is stimulated by testosterone. Testosterone also regulates the population of thromboxane A2 receptors thereby affecting platelet aggregation. Testosterone has a positive effect on the heart. Contrary to popular belief, high testosterone levels are positively correlated with low incidence of coronary disease. High levels of testosterone are associated with low amount of aortic plaque. Similarly low levels of testosterone have been shown to decrease high density lipoprotein-cholesterol (HDL-C or good cholesterol), increase low density lipoprotein-cholesterol (LDL-C or bad cholesterol), increase total cholesterol and increase triglycerides. Levels of testosterone are inversely proportional to presence of hypertension. It is believed that testosterone is a vasodilator (relaxes blood vessels) and therefore reduces blood pressure. Administration of testosterone has been shown to improve symptoms of angina and exercise tests. There is some evidence that testosterone may prove useful in the management of cardiac failure, but more research is needed.

Muscle, Fat and Bone

Testosterone is said to increase neurotransmitters, which in turn encourages tissue growth. Testosterone interacts with DNA receptors resulting in protein synthesis. Increase in growth hormone is also attributed to testosterone. So when we exercise, we can experience muscle growth.

Osteoporosis is lack of bone density and strength. Osteoporosis can lead to bone fractures, most commonly of the hip, vertebrae and forearm. Testosterone levels in young males correlate positively with bone size and peak bone mass. Testosterone increases bone density and stimulates the bone marrow to manufacture more red blood cells.

Total and free testosterone levels are inversely correlated with waist circumference and central (visceral) obesity rather than general obesity. Visceral obesity is a predisposition to metabolic syndrome, diabetes and cardiovascular disease. Increase in testosterone results in a decrease in fat mass and an increase in fat free mass as well as a decrease in central obesity. Testosterone helps in fat burning as well.

Metabolic Syndrome and Type 2 Diabetes

Metabolic syndrome is a combination of insulin resistance and/or impaired glucose function, abnormal lipid profile, and hypertension. Insulin resistance is a condition in which more and more insulin is required to keep normal glucose levels. It can lead to type 2 diabetes. Studies in non-diabetic men have shown that high levels of testosterone are correlated with low levels of glucose and insulin. Similarly low testosterone levels can predict future development of metabolic syndrome. High levels of testosterone have shown a decrease in insulin resistance and a decrease in glycosylated hemoglobin (HbA1c). The relationship between testosterone and metabolic syndrome is explained by the hypogonadal-obesity-adipocytokine cycle. Increase in body fat leads to an increase in aromatase levels, increase in insulin resistance, abnormal lipid profiles and increased leptin levels. Increase in aromatase levels leads to increase in conversion of testosterone to estrogen. Hence the testosterone levels decrease and visceral fat increases.

Increase in leptin levels act on the pituitary to decrease gonadotropin releasing hormone (GRH) levels which reduce luteinizing hormone (LH) and follicle stimulating hormone levels and subsequently reduces testosterone levels causing a vicious circle.

Hair

Testosterone is responsible for hair growth on the face, arms, chest, legs, armpit and genitals. When testosterone levels decrease, there is a loss of hair.

Cognitive Function and Mood

From childhood, males and females have differing cognitive function. This difference is attributed to sex hormones. Girls have stronger verbal skills while boys have stronger spatial skills. Testosterone levels show a positive correlation between future cognitive abilities and rate of cognitive decline. Testosterone seems to improve spatial cognition, verbal memory and working memory. Studies have shown that testosterone elevates mood and imparts a general sense of well-being. This seems to relate to the vasodilating action of testosterone which results in improved perfusion of the brain. Depressed aging men show low levels of testosterone and treatment with testosterone improves depression scores. However, the findings are conflicting.

Longevity

So far there have been four separate studies that have looked at the relationship between testosterone levels and longevity. In one study of 858 men above age forty, those with low levels of testosterone (< 250 ng/dL) had a 75%

higher mortality rate. Two studies in older patients found that those with low levels of testosterone died at a higher rate than those with high levels. Well designed studies need to be carried out in order to say this definitively. Also these studies only looked at natural levels of testosterone. Studies have not looked at the relation between increasing testosterone levels and longevity

Abuse

Athletes use testosterone and other anabolic steroids to improve their performance. These steroids can enhance muscle development, strength and endurance. The steroids increase protein synthesis in the muscle and therefore the muscles become larger and repair faster. However, this is considered abuse or doping. Lyle Alzado was a defensive end in the National Football League. He admitted to using anabolic steroids when he was playing professionally. But he also thought that the brain tumor that eventually led to his death was caused by steroid use. But later a scientist at the University of Wisconsin disproved this theory. Alzado also said that he suffered from steroid rage and that it made him very aggressive on and off the field. In 1988, Ben

Johnson, who won Olympic gold medal was stripped of the medal after it was discovered that he used anabolic steroids. In the 1990's baseball player Mark McGwire disclosed that he routinely used androstenedione (a supplement that gets converted to testosterone). Similarly, baseball player Barry Bonds was also suspected of using steroids to improve his play. Abuse is also reported to be widespread in Olympic athletes, wrestlers, mixed martial arts fighters, bicyclists and law enforcement personnel. The Nevada Athletic Commission has banned the medical use of testosterone therapy.

A urine test to check different testosterone ratios has been used to detect abuse. Hair analysis has also been used to detect abuse.

CAUSES OF LOW TESTOSTERONE

Testosterone to me is so important for a sense of well-being when you get older.

Sylvester Stallone

As we age, there is a natural tendency to lose testosterone. But other factors such as obesity, high blood sugar and exposure to environmental pollutants can also cause low testosterone. But it is important to recognize the difference between low testosterone due to aging and a medical condition called hypogonadism. It has also been referred to as andropause, androgen deficiency or late onset hypogonadism.

Diagnosis

Diagnosis of medical or clinical low testosterone (hypogonadism) is done by checking the levels of total testosterone in blood. It is recommended to take the blood first thing in the morning because testosterone levels are highest at that time. They drop by 30-40% by mid-afternoon. The test should also be repeated on different days and times to get a good idea of inherent variability and to prevent a false positive result. As shown in Figure 6, total and free testosterone decrease with age and total testosterone is comprised of testosterone bound to proteins (which is not bioavailable) and free testosterone (which is bioavailable). About 45% testosterone is bound to sex hormone binding globulin (SHBG). This is a tight bond and renders the testosterone inactive. About 50% is bound to albumin, but it is loosely bound and is considered bioavailable. Finally 2-3% testosterone is free. It is believed that decrease in free testosterone has a greater effect than decrease in total testosterone. If the total testosterone level is less than or equal to 300 nanogram per deciliter (ng/dL), then hypogonadism is diagnosed.

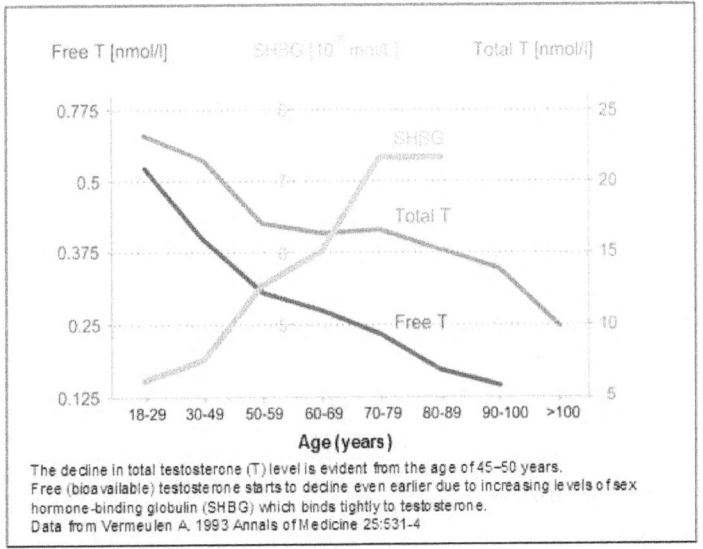

The decline in total testosterone (T) level is evident from the age of 45–50 years. Free (bioavailable) testosterone starts to decline even earlier due to increasing levels of sex hormone-binding globulin (SHBG) which binds tightly to testosterone. Data from Vermeulen A. 1993 Annals of Medicine 25:531-4

Figure 6: Relationship between total, free and bound testosterone

As for the levels of free testosterone, normal range is 150-250 picogram per deciliter (pg/dL). There is another test that measures free testosterone. It is called equilibrium dialysis. Not every laboratory is capable of doing this test and it is also expensive. If that test shows that free testosterone is less than 150 pg/dL, it may be hypogonadism. It should be noted that experts do not agree on what constitutes "low" testosterone. Also the levels at which symptoms appear are different for different people.

Another diagnostic test is a qualitative test. It is called Androgen Deficiency in Aging Males or the ADAM test. It was designed by Dr. Morley, the director of endocrinology and geriatrics at the St. Louis University School of Medicine. It consists of following 10 questions:

1. Do you have reduced libido (sex drive)?
2. Do you have a lack of energy?
3. Do you have a decrease in strength and endurance?
4. Have you lost height?
5. Have you noticed less enjoyment of life?
6. Are you sad and/or grumpy?
7. Are your erections less strong?
8. Have you noticed a recent deterioration in your ability to play sports?
9. Are you falling asleep after dinner?
10. Is your work performance suffering?

The responses are either a Yes or a No and based on the response, a suggestion is made. But these questions are such that anyone could be experiencing one or more of those "symptoms" unrelated to low testosterone. In other words, anyone could fail the test and go to their doctor asking for testosterone replacement therapy (TRT). These symptoms are also difficult to measure.

Signs and Symptoms

The three main signs of hypogonadism are decrease in muscle mass and strength, decrease in bone mass and osteoporosis and increase in body fat around the middle of the body. Since testosterone stimulates red blood cell production in the bone marrow, another sign is anemia. It is estimated that about 20% of the men above 50 years of age may have low testosterone levels, but are asymptomatic. The symptoms can be divided into sexual and non-sexual. Sexual symptoms are loss of libido, lack of sexual desire, difficulty achieving orgasm, decreased intensity of orgasm, reduced amount of ejaculatory fluid and erectile dysfunction (ED). It should be noted that ED can be caused by problems with blood flow to the penis, certain medications and surgery. The non-sexual symptoms are low energy/increased fatigue, lack of motivation, sleeplessness, and forgetfulness. But as mentioned before, they are difficult to measure and distinguish from symptoms associated with aging.

Depression can also be a cause of lack of sexual desire or loss of libido, therefore a blood test should be done prior to starting testosterone treatment.

Aging

As we age, the testosterone levels gradually decrease about 1% every year after age 30 (see Figure 1). Therefore incidence of hypogonadism also increases with age as shown in Figure 7

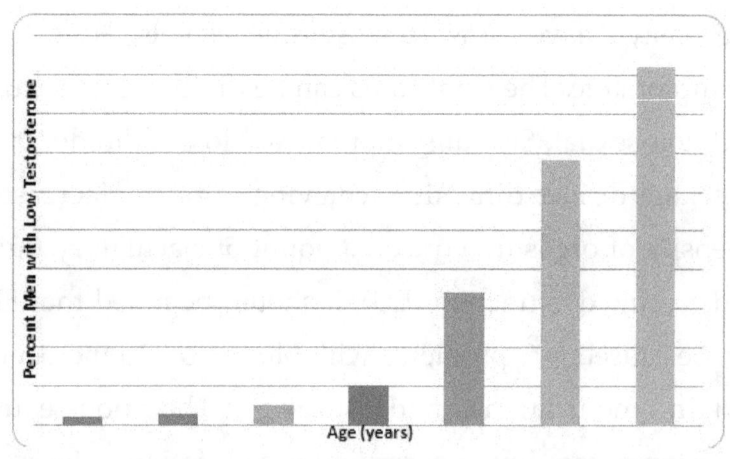

Figure 7: Age related increase in men with low testosterone

Diseases

Hypogonadism can be divided into two types:

1. Primary- It is also called primary testicular failure and as the name suggests, this type is caused by some problem with the testicles.
2. Secondary- This type is caused by a problem with the hypothalamus-pituitary axis.

Either type of hypogonadism can be caused by an inherited genetic defect or a problem or ailment that occurs later in life such as a disease or accident. Given below are the causes of both.

Primary hypogonadism

Common causes of primary hypogonadism include:

Undescended testicles
The testicles develop inside the abdomen before birth and then move down into their permanent place in the scrotum. Sometimes one or both of the testicles may not be

descended at birth. This is a self-correcting condition in early years and seldom requires treatment. However, if it does not get corrected naturally in early childhood, it may cause malfunction of the testicles and reduced production of testosterone.

Leydig Cell Aplasia

It is a very rare genetic disorder which affects 1 in 1 million males. It is caused by the inability of the Leydig Cells to respond to luteinizing hormone. As described before, luteinizing hormone signals the Leydig Cells in the testicles to secrete testosterone. It is also sometimes referred to as luteinizing hormone insensitivity.

Injury to the testicles

The testicles are located outside the abdomen and are not protected. Thus they are susceptible to injury. Damage to normally developed testicles can cause hypogonadism. On the other hand if only one testicle is damaged, it may not interfere with total testosterone production.

Klinefelter syndrome

This is a genetic disorder involving abnormality of the X and Y chromosomes. A male normally has one X and one Y chromosome. But in Klinefelter syndrome, in addition to a Y chromosome, two or more X chromosomes are present. The genetic material in the Y chromosome determines the sex of a child and related development. The extra X chromosome results in abnormal development of the testicles, which in turn results in underproduction of testosterone.

Mumps

In addition to a mumps infection of the salivary glands, if there is testicular mumps infection during adolescence or adulthood, long-term testicular damage may occur. This may affect normal testicular function and testosterone production.

Hemochromatosis

When there is too much iron in the blood, the condition is called hemochromatosis. It can cause pituitary gland dysfunction or testicular failure affecting testosterone production.

Cancer and Cancer treatment

Presence of testicular cancer can directly impact the level of testosterone. However, other cancers/tumors can also affect testosterone levels. Pituitary tumors also affect testosterone secretion from the testicles. Common cancer treatments like chemotherapy or radiation therapy can impair testosterone and sperm production. Although the effects of both treatments are often temporary, permanent infertility may occur. But many men regain their fertility within a few months after treatment ends. If need be, sperm can be preserved before starting cancer therapy.

Secondary hypogonadism

Secondary hypogonadism can be caused by a defect of the hypothalamus-pituitary axis. The testicles are normal. Many conditions that can cause secondary hypogonadism are described below:

Pituitary disorders

Any abnormality in the pituitary gland such as a pituitary tumor or a brain tumor near the pituitary can disrupt the release of hormones from the pituitary gland to the testicles, affecting normal testosterone production. In addition, surgery or radiation therapy used to treat the tumor, may impair pituitary function and cause hypogonadism.

Kallmann syndrome

As described earlier, the hypothalamus is a part of the brain that controls the secretion of pituitary hormones. Kallmann syndrome is characterized by an abnormality of the hypothalamus that affects hormone secretion causing hypogonadism.

Inflammatory disease

Some inflammatory diseases, such as sarcoidosis (growth of a group of inflammatory cells or granulomas on different organs, predominantly lung and lymph nodes), histiocytosis (a group of syndromes involving an abnormal increase in immune cells or histiocytes) and tuberculosis (a disease affecting the lungs, but also other organs), involve the hypothalamus and pituitary gland and can affect testosterone production, causing hypogonadism.

HIV/AIDS

HIV/AIDS can cause low levels of testosterone by affecting the hypothalamus, the pituitary and the testes. In a study, HIV positive men had approximately the same total testosterone as HIV negative men, but they had lower levels of free testosterone. HIV positive men also have higher levels of Sex Hormone Binding Globulin (SHBG), which binds to testosterone. This explains why HIV positive men have lower levels of free testosterone. Men in advanced stages of HIV can have inflammation of the testicles, causing them to produce low amounts of testosterone. Some anti-retroviral treatments (ART) can increase the amount of body fat which also affects testosterone levels.

Tumor necrosis factor (TNF) and interleukin-1 (IL-1) are regulated up in HIV patients and can cause reduction of testosterone levels.

Medications

The use of certain drugs, such as opiate pain medications and some hormones, can affect testosterone production.

Obesity

Lifestyle factors like being significantly overweight at any age may be linked to hypogonadism. This is because body fat helps convert testosterone to estrogen.

Lack of Sleep

Another lifestyle factor is sleep or lack thereof. A 2011 study in the Journal of American Medical Association found that sleep deprivation can cause a 10-15% reduction in testosterone levels. Lack of sleep can also cause erectile dysfunction.

Normal aging

As has been discussed earlier, older men generally have lower testosterone levels than younger men do. As men age, there's a slow and continuous decrease in testosterone production.

Concurrent illness

The physical stress of an illness or surgery, as well as significant emotional stress can temporarily inhibit testosterone production. This is a result of diminished signals from the hypothalamus and usually resolves with successful treatment of the underlying condition.

Kidney Disease
People suffering from End Stage Renal Disease (ESRD) can have hypogonadism.

Other Factors
A study by Northwestern University in 2011 showed that fatherhood lowers a man's testosterone. The study which lasted almost 5 years, followed 624 males in their early to mid-20's before and after they became fathers. Some doctors suggest that demands of fatherhood result in lack of sleep and lack of exercise consequently lowering testosterone.

A 2007 study carried out by New England Research Institute and reported in the Journal of Clinical Endocrinology and Metabolism showed that men today have low median testosterone levels than men 20 years ago (see Figure 8).

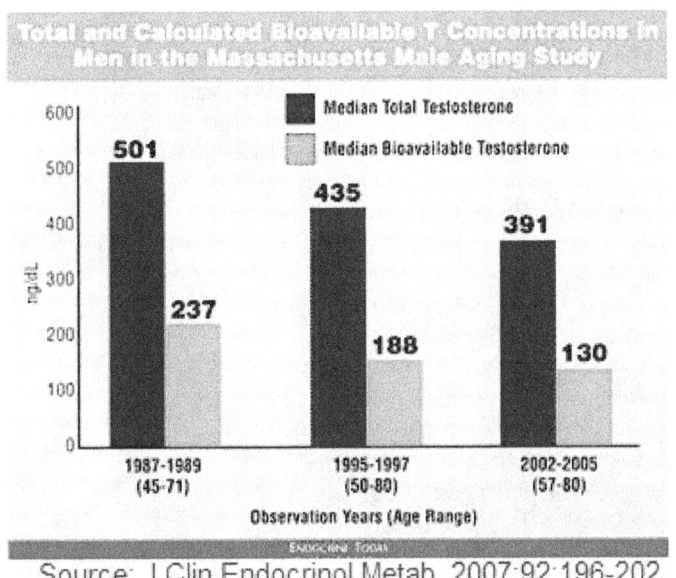

Source: J Clin Endocrinol Metab. 2007;92:196-202.

Figure 8: Comparison of median testosterone levels in men from different time periods

They said that these declines cannot be explained by ageing and lifestyle and that other factors may be responsible. Environmental pollutants can also affect testosterone levels. A series of studies carried out by researchers at the University of California at Berkeley showed that atrazine, a common fertilizer, can convert male frogs into female frogs. Other studies in fish, amphibians, birds, reptiles, laboratory animals and human cell lines have shown that atrazine levels as low as parts per billion can interfere with testosterone and estrogen. Last but not least factors such as smoking, alcohol and drug abuse, diabetes, vasectomy and pollution can also affect testosterone levels.

PHARMACEUTICAL TREATMENTS

Years ago I was diagnosed with a condition,
and my doctors prescribed human growth hormone and
testosterone for its treatment. Under medical supervision,
I have continued to use both medications.

Sylvester Stallone

One in 5 million men in the US have been diagnosed with low testosterone. But a study in International Journal of Clinical Practice seems to indicate that there are about 12.5 million men in the US that have low testosterone, but are not receiving treatment for it. In 2012, $100 million were spent on advertising for testosterone replacement products and the gross sales that year were $2.2 billion. Indeed a study in the Journal of American Medical Association (JAMA) Internal Medicine found that the number of prescriptions for testosterone from 2001 to 2011 in men 40 and older more than tripled. It should be noted that large scale studies on long term effects of testosterone replacement therapy have not been carried out. Some studies that have been performed have yielded mixed results. So if in fact after a blood test has confirmed the diagnosis of hypogonadism and you have the three main symptoms of decrease in muscle mass and strength, decrease in libido and fat around your mid-section, treatment should be started. According to the US Food and Drug Administration (US FDA), testosterone replacement therapy (TRT) is only approved for adult males suffering from "a medical condition associated with a deficiency or absence of endogenous testosterone (primary or secondary hypogonadism). There are no testosterone drugs approved as a treatment for low testosterone levels, often referred to as 'lowT', without an associated medical condition." But doctors prescribe these treatments for "off label" use. It should also be noted that many doctors do not agree on whether treatment is justified. The main reason for the disagreement is that large, long-term studies looking at the risks and benefits of testosterone replacement therapy have not been done. According to a 2004 report, 31 placebo-controlled trials have been carried out for TRT. The largest of these trials had 108 patients. Only 1 of those trials lasted beyond 1 year and 25 trials lasted 6 months or less. Many studies have provided mixed results. In the February 2016 issue of the New England Journal of Medicine, results from a multi-center, randomized, double-blind, placebo-controlled trial were published. This study was funded by the National Institutes of Health and the objective was to address issues raised by the Institute of Medicine (IOM). The IOM thought that there was a dearth of randomized, double-blind, placebo-controlled clinical trials involving older men and that there was a lack of evidence supporting the benefits of testosterone therapy. The study consisted of 790 men who were 65 years or older and a testosterone level of less than 275 ng/dL. The trial was supposed to measure seven different outcomes: sexual function, vitality, physical function, cognitive function, anemia, bone density and cardiovascular status. However, the primary outcomes were sexual function, vitality and physical function. The interesting fact about this study was that 51,085 men volunteered, but only 1.5% (790 men) actually had low testosterone. This implies that the majority of the men did not know what their testosterone levels were. Anyway, the results showed that sexual function, sexual desire and erectile function only improved modestly over the placebo group. There were no improvements in vitality or physical function as compared to placebo. However, the study did not answer the question whether testosterone therapy is safe. The study also did not resolve whether middle aged men (who are the biggest group of people taking testosterone) can really benefit, neither does it say whether those men whose testosterone level is slightly above 275 ng/dL would improve. The study does show that for a specific group of men (65 years or older and a testosterone level of less than 275 ng/dL) testosterone therapy has modest benefits.

One group of doctors believes that since other symptoms of aging such as poor vision, poor hearing and joint pain are treated then why not treat low testosterone? They feel that instead of reacting and managing a disease, doctors need to be proactive and prevent disease. Also in today's day and age, there is a shift from evidence based medicine (which is medical decisions based on evidence gathered from a group of people) to personalized medicine (medicine tailored towards what an individual is experiencing). People are living longer. Today the life expectancy in the US is 78 years as compared to 69 years in 1965. They want to lead more active lives and still be able to do things that they used to do when they were younger. Yet other health experts feel that use of testosterone therapies is outpacing the research into its efficacy and side-effects. They feel that TRT is where hormone replacement therapy (HRT) for women was 20 years ago. At that time post menopausal women were prescribed hormone replacement therapy to treat some of the signs and symptoms of menopause such as hot flashes, osteoporosis and weight gain. There was a surge in HRT prescriptions. But as more research was done and severe side effects were observed, a downward trend followed.

Hype v/s Fact

As far back as 1st century Rome and the medieval times, there are accounts of remedies for impotence and erectile dysfunction. However, it was not until the 1920's and 30's when emergence of endocrinology determined that hormones play a role in producing the sexual traits of men and women. The transformation of neutered chickens (capons) into roosters after being injected with testosterone cemented its role as the hormone responsible for male characteristics.

Older men experience lack of sexual desire and loss of libido. Since testosterone gradually decreases in aging, it was not long before a connection was made between testosterone and lack of sexual desire and loss of libido. All of a sudden it looked like testosterone is the fountain of youth. Once testosterone therapy is started, other effects such as improved muscle mass and strength, loss of fat, increase in endurance are also observed, which further establishes testosterone as the magic cure for aging. Direct to consumer advertising is asking men if they feel "less of a man" or a "shadow of their old self". The advertisement shows the benefits of TRT and encourages the viewer to speak to their doctor. This type of advertising is called unbranded advertising and it is not scrutinized by the FDA. According to Kantar Media, advertising costs for testosterone treatment was $14 million in 2011. Many male baby boomers that are now getting older are flocking to these types of products hoping to make their lives better. Indeed this myth is being propagated by Low T centers sprouting up like weeds all over the country. But critics think this is just another way to trick healthy men into thinking that they need TRT. They fear that men whose problems are not related to low testosterone are also taking them and potentially exposing themselves to the risk of side effects. Indeed a recent study in JAMA found that about 25% of the men received the prescription without a blood test.

Testosterone Replacement Therapy

Testosterone Replacement Therapy (TRT) consists of taking synthetic testosterone. Testosterone, if administered orally, is degraded in the stomach. But if some chemical groups are added to testosterone (example methyl group), then it can be administered orally. A review of FDA approved treatments for testosterone indicates that most oral treatments have been discontinued. Oral forms can also cause liver toxicity, benign and malignant cancer, increase in low density lipoprotein-cholesterol (LDL-C) and decrease in high density lipoprotein cholesterol (HDL-C). One oral form containing testosterone undecanoate is available outside the US. So all of the TRT treatments are administered non-orally. I will list them below:

Injectable Testosterone

Injected testosterone lasts only for a few hours in the blood. In order to increase the amount of time it stays in the blood, testosterone is coupled with a chain of carbon and hydrogen atoms called ester. Testosterone is water soluble and by adding an ester chain, it becomes less water soluble and more fat soluble. The longer the ester chain, the higher the fat solubility and lower the water solubility. Testosterone ester is accumulated in the fat tissue and is slowly released. In other words it acts like a depot. The released testosterone ester is converted to free testosterone by enzymes called esterases. There are many injectable forms of testosterone, most common being testosterone cypionate and testosterone enanthate. They come in 100 milligram per milliliter (mg/mL), 200 mg/mL or 300 mg/mL strength. Depending upon the dose (100, 200 or 300 mg/mL) they are administered once a week, once every two weeks or once every three weeks respectively. There is an increase in testosterone levels immediately after the injection has been administered and it could cause some side effects. On the other side, as you near the end of treatment period, there is a drop in testosterone levels and one may experience symptoms before the next dose. The optimal dose seems to be 200 mg/mL, but that can only be determined by the doctor. Generally a blood test is done mid-cycle to determine the testosterone level and adjust frequency/dosage accordingly. There is a long acting injectable that is dosed every ten weeks, but it is not yet available in the US. These treatments are administered as an intra-muscular (IM) injection into the buttocks and the cost is about $40.00 to $200.00 per month. Usually the treatment is administered in a doctor's office, but men can be trained to self-administer. Self-administration can be done on the thigh.

Topical Testosterone

Topical testosterone treatments are applied to the skin. Steroid hormones are readily absorbed through the skin and that is why this route is attractive. They come in a gel form or a transdermal patch form (like the nicotine or motion sickness patches). Unlike the injectable treatments, these treatments are administered daily in low doses. About 12% of the market for testosterone treatment is in the form of a transdermal patch. The patch is applied to the lower abdomen. The patch comes in 2 mg per 24 hours or 4 mg per 24 hours. The patches can cause skin irritation and red blotches after removal. Also some men might be uncomfortable wearing the patch in a public place such as a pool or gym. Gels constitute 60% of the testosterone market. The gels come in a concentration of 1.62% or 20.25 mg per 1.25 gram actuation. Some treatments are 30 mg per 1.5 mL. The gel is applied to upper arms, shoulder or chest. Showering is not recommended for up to two hours after applying the gel. Testosterone from gels is better absorbed than from patches. But there is a lot of person-to-person variability. About 15-20% of the men absorb the gel poorly or not at all. For these men, injection is the best option. Another option is to switch to another gel. Also, earlier in the treatment period, blood tests need to be done to determine if the testosterone levels are raised. The advantage of topical treatments is that since they are administered daily and the dose is low, the side effects are typically low or none. They are less cyclical compared to the injection. But they are costlier than the injection. Typical cost of topical treatments is $200 to $500 per month. There are pharmacies that compound the cream, but compounding pharmacies are not regulated by the FDA.

As with any other treatment, if a cream from the pharmacy is being used, blood tests should be carried out to make sure the cream is working.

Other type of topical treatment is the buccal tablet. These tablets stick to your gums and the dose is 30 mg. Their side effect profile and cost are similar to the gels and patches. It is difficult to maintain the patch on the gums.

Subdermal Testosterone

This type of treatment is implanted under the skin in the form of pellets. The insertion is done at the doctor's office with local anesthesia. This type of treatment delivers testosterone over 4 to 7 months. The standard dose is four pellets of 200 mg (800 mg). These treatments have not been as extensively studied as other treatments.

Nasal Spray

In May 2014, FDA approved a nasal spray for testosterone. Each spray contains 5.5 milligram (mg) of testosterone. The dose is two sprays (one per nostril) for a total dose of 33 mg daily.

Suppositories

Suppositories are bullet shaped objects made from wax base. They are inserted in the rectum. Testosterone suppositories are available in Europe, but not very popular in the US. However, compounding pharmacies can manufacture them. The advantages of the suppositories are that they bypass the stomach and liver so they are absorbed faster. Disadvantages are the route of administration and sudden burst in testosterone levels followed by a drop requiring higher dose frequency.

Non-Testosterone Drugs

Human chorionic gonadotropin (hCG)

It is a natural hormone made in the placenta of pregnant women. It is found in their urine. The purified form is available as an injectable treatment. It is available as a powder that has to be mixed with sterile water or is premixed. It is said to mimic the action of luteinizing hormone, which signals the Leydig cells to secrete testosterone. The disadvantage of testosterone drugs is that they tend to suppress endogenous secretion of testosterone which in turn can lead to lowering of sperm count causing temporary infertility. However, since hCG is not testosterone, it does not suppress endogenous testosterone.

Human Growth Hormone (HGH)

Human growth hormone does not directly affect testosterone levels. Similar to testosterone, it is in high amounts during childhood and declines as we grow older. But a study by Dr. Rudman in the New England Journal of Medicine showed that HGH helps build lean muscle, reduce abdominal fat and increase bone density in older men. So it can act synergistically with TRT to improve overall health. Side effects of HGH include acromegaly or enlargement of hands and feet.

Clomiphene

Another drug that is used to treat low testosterone is clomiphene. It also works on the hypothalamus-pituitary axis and inhibits the negative feedback loop for LH and FSH (see Figure 4). It is not a steroid and it is cheaper and does not have some of the side effects of testosterone. It is available in a tablet form, so patient compliance is better. It is not FDA approved to treat hypogonadism, although it is used off-label by doctors. Another form of clomiphene called enclomiphene is undergoing clinical trials to get FDA approval for treating hypogonadism.

Aromatase Inhibitors

Aromatase is an enzyme that converts testosterone to estrogen. A class of drugs called aromatase inhibitors have been used to increase the levels of testosterone. Two drugs in this class are anastrozole and letrozole. Letrozole is more potent than anastrozole. These drugs also do not suppress endogenous testosterone and sperm production is not affected.

Testosterone Mechanism of Action

In younger males, endogenous testosterone and dihydrotestosterone (DHT) are responsible for the normal growth and development of prostate, seminal vesicles, penis and scrotum. They are also responsible for maintenance of secondary sex characteristics such as facial, pubic, chest and axillary hair; laryngeal enlargement, vocal cord thickening, alterations in body musculature and fat distribution. Testosterone and DHT bind to receptors and this complex undergoes a structural change that allows it to enter the cell nucleus. In the nucleus testosterone and DHT bind to areas on deoxyribonucleic acid (DNA) to form specific proteins. In collaboration with other DNA binding proteins, testosterone and DHT can stimulate muscle growth. A second mechanism is that testosterone gets converted to estradiol and is responsible for strengthening the bones. In the central nervous system, testosterone is converted to estradiol which provides the negative feedback loop to the hypothalamus (see Figure 4). In hypogonadism patients, externally administered testosterone reduces fat and increases muscle mass and strength. It also increases the length of type I and type II muscle fibers.

Benefits of TRT

For men suffering from hypogonadism, TRT has proven benefits. Positive effects are seen in muscle mass and strength, sexual desire and libido, lowering of fat, increase in bone mineral density, cognition, cardiovascular effects and a general sense of well being.

Sexual Desire and Function

Levels of free testosterone in the blood have been correlated with erectile function. There are other causes of erectile dysfunction, but in men who are suffering from hypogonadism, TRT seems to improve erectile function. TRT also enhances libido and sexual function.

Bone Mineral Density

In hypogonadal men, the incidence of osteopenia, osteoporosis, and bone fracture is higher than normal men. Bone mineral density in hypogonadal men increases with TRT. Studies have also shown positive effects on lumbar spine bone density.

Improved Body Composition

TRT had a positive effect on reducing fat mass and increasing lean body mass. There was also a positive correlation between TRT and increase in muscle mass and strength.

Cognitive Function

Low levels of testosterone have been shown to increase the levels of Alzheimer's Disease (AD) related peptides. Small studies of short duration have shown some improvement in AD symptoms. The reasoning is that testosterone reduces the formation of amyloid plaque, the main cause of AD. Age related decrease in levels of testosterone correlates with loss of visual and verbal memory. There is also a relationship between low levels of testosterone and mathematical reasoning. TRT seems to improve cognitive function.

Depression

In a study at McLean Hospital in Boston, half the patients who were depressed and on medication showed borderline to low levels of testosterone. All patients who received a testosterone supplement in addition to depression medicine showed greater improvement in mood. However, the improvement was not to the same degree. Some patients did extremely well, while others did not.

Metabolic Syndrome, Type 2 Diabetes and Cardiovascular

Metabolic syndrome is characterized by obesity, hypertension, dyslipidemia, impaired glucose regulation and insulin resistance. Obesity decreases free and bound testosterone levels. TRT has a positive correlation with reduction in the characteristics of metabolic syndrome. Higher levels of testosterone have cardio-protective effect. There is a weak direct correlation between diabetes and testosterone levels. However, indirectly, since testosterone increases lean body mass, improves metabolism and reduces fat, it has a positive effect on insulin modulation.

Anemia

Testosterone stimulates erythropoiesis (production of red blood cells), blood hemoglobin and erythropoeisis in the bone marrow. This is a direct effect of testosterone on renal production of erythropoietin. Hypogonadal men have 10% to 20% decrease in hemoglobin resulting in anemia.

HIV/AIDS

As mentioned previously, testosterone levels are decreased in men with HIV/AIDS. Testosterone treatment improves lean body mass, strength and libido in men with HIV/AIDS for up to six months. However, effects of long term treatment have not been studied

Side Effects of TRT

The side effects of TRT are affected by age, lifestyle and other concurrent medical conditions. Some of the common side effects seen with all TRT treatments are injection site/application site pain, redness, swelling, itching, burning sensation. Nausea, vomiting, change in skin color, hair loss, tiredness and lack of sleep are other side effects. For subdermal treatments there is a risk of infection. Patches and gels can cause rash and redness of skin. For the nasal spray the side effects are nasal irritation, runny nose, congestion, nose bleeding and nasal dryness. But these treatments also come with serious side effects listed below:

Benign Prostatic Hyperplasia and Prostate Cancer

Increase in testosterone results in an increase in prostatic volume (enlarged prostate) and also an increase in Prostate Specific Antigen (PSA). The enlargement of the prostate can cause urinary problems and result in frequent urgent urination. So far there is no correlation between TRT and prostate cancer, however when prostate cancer is pre-existing, TRT can accelerate the growth of tumors. Therefore, hypogonadal men should be screened for prostate cancer before starting TRT. The relationship between prostate cancer and testosterone was based on a study that was first performed in 1941. But that study was done on four men and only one of them showed growth in prostate cancer. In addition to the small number of patients, the study used a test which was the state of the art at that time, but in later studies has been shown to be unreliable. However, this knowledge was propagated. Another study in the 1980's used fifty two patients and forty five patients showed signs of prostate events. But four patients actually showed improvement. Many studies since then found that low testosterone, not high testosterone is responsible for prostate cancer. Yet another study found that TRT increases the levels of free and bound testosterone in the blood, but testosterone levels in the prostate remain unchanged. It was also observed that testosterone receptors in the prostate get saturated at low levels and any excess testosterone really doesn't make a difference. Finally, in 2008, a study was carried out that combined data from eighteen separate studies consisting of 9000 patients. One third of those patients had prostate cancer and the remaining did not. The study found no relationship between testosterone levels and prostate cancer. In summary, the current state of affairs is that TRT does not increase the risk of prostate cancer, high levels of testosterone don't lead to prostate cancer, low levels of testosterone don't protect against prostate cancer, in fact the opposite maybe true. Clinicians estimate that in order to determine if testosterone therapy causes prostate cancer, a study consisting of 5,000 patients lasting 3-5 years is necessary and a review of current and planned clinical studies shows no such study being planned.

Cardiovascular

Until 2010, studies on TRT had not resulted in the death of a patient due to cardiovascular events. But as more studies were done in larger populations, a 30% increase in deaths due to cardiovascular events has been reported. A study in 2013 was stopped due to a disproportionate increase in adverse events in TRT group versus placebo. A study in 2014 had to be stopped because of death due to cardiovascular events. In 2014 the FDA required all TRT to put a warning on the label for possible cardiovascular adverse events. TRT has also been linked to a risk of mini stroke, stroke or blood clots. Cardiovascular side effects are caused by several different mechanisms. One mechanism is the increase in thromboxane A2 and platelet aggregation. Platelets play a role in coronary plaque formation. If this plaque ruptures, it could cause acute coronary syndrome.

Dihydrotestosterone, which is a metabolite of testosterone, increases smooth muscle proliferation, consequently causing monocyte activation. Monocytes also play a role in arterial plaque formation.

Finally, testosterone worsens the effect of sleep apnea and sleep apnea can also cause atherosclerosis.

People with existing high blood pressure can see an increase in blood pressure due to TRT. It is related to fluid retention (see below) and does return to normal in most cases.

The risk of cardiovascular events prompted the FDA to have an advisory committee meeting to discuss TRT. The committee overwhelmingly voted to use TRT in patients with hypogonadism rather than people with age related low testosterone. As with prostate cancer, the connection between heart attacks and testosterone was first reported when a body builder in his 30's suddenly died of a heart attack and it was found later that he was taking steroids. Other studies have linked testosterone to increase in "bad" or LDL-cholesterol. But that has also been proven wrong.

Another concern was that testosterone can cause atherosclerosis. However, some studies have shown that testosterone in fact reduces atherosclerosis. The reason for these conflicting findings is how the studies are carried out and condition of the patients in the study. If the studies are not properly designed or if the number of patients is small the findings from such studies should be taken with a grain of salt. The subject of clinical trials and phases of clinical development has been covered in detail in the book Taming Ebola (please link to Amazon Kindle listing)

Liver Problems

There have been reports of benign and malignant hepatic tumors, hepatotoxicity and liver failure, associated with oral TRT. Other problems reported are hepatocellular adenoma and carcinoma. Liver toxicity is predominantly seen with oral testosterone therapy and not with non-oral treatments.

Cognitive Function

As mentioned above low testosterone can have a negative effect on cognitive function. But since TRT increases hemoglobin and red blood cells, high concentration of both are associated with reduction of cognitive function. Elevated hemoglobin causes storage of unhealthy levels of iron which in turn can cause dementia and oxidative damage. High levels of hemoglobin and red blood cells also increases the risk of Alzheimer's Disease.

Sleep Apnea

Sleep apnea is a condition in which airflow is interrupted or stopped during sleep due to narrowed or blocked airway. It can cause atrial fibrillation or cardiac ischemia (not enough blood flowing to the heart). According to a 2008 study published in Clinical Endocrinology, TRT can induce or exacerbate sleep apnea. The exact mechanism is not known, but is believed to be related to the effect of testosterone on the central nervous system.

Polycythemia

Since testosterone stimulates red blood cell production, it can increase the amount of red blood cells causing clots.

Hormonal Disruption

TRT can cause hormonal disruption resulting in development of oily skin, acne and enlargement of breast in men. Studies have also shown the external administration of testosterone suppresses endogenous secretion of testosterone causing lowering of sperm production. Shrinkage of testicles has also been observed.

Fluid Retention

TRT can cause fluid retention and should not be prescribed to people with heart failure, kidney disease and liver damage. These effects are seen immediately commencing the treatment, but fluid levels do return to normal. Occasionally, swollen feet and legs are seen.

Risk to Friends and Family

Friends and family of the person undergoing TRT, especially women and children, are not supposed to touch the medicine. Skin contact can also be risky, specifically with gels. There is a risk that women may develop male like characteristics and children may show premature puberty, enlarged sex organs and aggressive behavior.

Contraindications

TRT is not advisable in patients with existing prostate or breast cancer, prostate nodules or unusually high levels of Prostate Specific Antigen (PSA). It is also contraindicated in patients with existing Benign Prostatic Hyperplasia, severe lower urinary tract symptoms, congestive heart failure and existing untreated sleep apnea. It is also contraindicated in Chronic Kidney Disease (CKD) patients. These patients already have a high cardiovascular risk.

Patient Monitoring

Prior to starting TRT a baseline determination of the following parameters should be performed:

1. Total and free testosterone
2. Prostate Specific Antigen
3. Hemoglobin
4. Hematocrit (ratio of red blood cell volume to total blood volume)
5. Lipid profiles (Total Cholesterol, LDL-C, HDL-C, Triglycerides)
6. Liver and kidney function test
7. Blood sugar and Hemoglobin A1C
8. Sleep apnea
9. Bone Mineral Density
10. Lower urinary tract symptoms

The same parameters should be tested at 3, 6, 12 months post treatment. Also the type of treatment (injectable versus topical) should be discussed. If post treatment parameters do not show improvement, discontinuation of TRT should be considered.

NATURAL TREATMENTS

Let thy food be thy medicine and thy medicine be thy food.

Hippocrates

As a general rule, the first treatment for any medical condition is diet and lifestyle changes. TRT is indicated only for people who have been diagnosed with hypogonadism. If there is an age related decrease in testosterone levels, there are other alternatives to pharmaceutical treatments.

Diet

Maintaining a healthy diet consisting of leafy green vegetables, fruits, whole grains, and legumes can help. Parsley is commonly used as a garnish and usually discarded before eating, but it is rich in apigenin. Apigenin lowers the levels of aromatase (the enzyme which converts testosterone to estrogen) thus increasing levels of testosterone. Also, blueberries and other berries contain flavonoids, which improve sexual function, especially erectile function.

As far back as 1967, a researcher in Egypt noticed that feeding raw onion juice to rats increased the size of their testicles. These findings were reproduced by a study in Iran in February 2010. The researchers fed raw onion juice to male rats for 20 days. The animals receiving onion juice showed increases in levels of testosterone, LH and FSH. The effect seems to be due to the presence of selenium compounds and phenols like quercetin and isothamnetin, which are antioxidants. These antioxidants bind free radicals that damage the cell membrane and form mutagenic substances like malondialdehyde. Antioxidants in onion juice neutralize free radicals in the testes.

Whenever possible eat organic food because pesticides are like estrogen, they affect testosterone levels. Similarly foods packaged in plastic containers and aluminum cans contain bisphenol A. Bisphenol A is said to mimic the effect of estrogen, thereby lowering testosterone. Increasing protein and reducing carbohydrate intake (1 carb: 2 protein ratio) also increases testosterone levels. Stay away from processed foods and refined sugar. Studies also seem to indicate that a diet containing 25% fat is beneficial for increasing testosterone levels. Ideally fat intake should comprise of polyunsaturated fats from foods like salmon, tuna, walnuts, pistachios, flax seeds and pumpkin seeds; monounsaturated fats from olive oil, avocados, and peanut butter; and saturated fats from coconut oil and animal sources of protein. Zinc deficiency is often observed in hypogonadal men. But there are plenty of foods that can provide zinc. For example, oysters, red meat, beans, nuts and whole grains. A diet rich in all those foods can increase testosterone levels.

A report by the Endocrine Society observed that glucose can decrease testosterone levels by 25%. So reducing dietary sugar intake can have a favorable outcome.

Testosterone levels decrease with age. The enzyme aromatase converts testosterone to estrogen. This causes an imbalance in the testosterone to estrogen ratio. Diet can also reduce estrogen and improve the ratio. Increasing dietary fiber can reduce estrogen. Fiber binds to bile acids and helps remove estrogen. Fiber is usually found in green leafy vegetables. Cruciferous vegetables such as broccoli, brussels sprouts, cauliflower and cabbage reduce the amount of stored estrogen. Dairy products include estrogen. So switching to non-dairy products can help.

Exercise

The type of exercise that most positively affects testosterone levels is resistance training. Resistance training is an exercise in which muscles contract against an opposing force for e.g. dumbbells or free weights. Resistance training can also increase strength. Strength training also increases testosterone levels. Strength training is using heavy weight and less repetition. Using more than one muscle group (compound exercises) also seems to increase testosterone levels. Finally high intensity training with intermittent fasting is also beneficial for increasing testosterone levels. High intensity training involves working hard for short periods of time like 30 seconds or one minute followed by a short rest and repeating the sequence. Aerobics or moderate prolonged exercise can actually lower testosterone. Intermittent fasting increases satiety hormones like insulin, glucagon like peptide-1, leptin and cholecystokinin which in turn increase testosterone levels. Exercise (especially resistance training) frees up receptors so human growth hormone (HGH) can bind to those sites. Exercise can improve your mood by stimulating the brain to secrete chemicals that make you feel happy.

Lifestyle Changes

As has been mentioned earlier, obesity can result in low testosterone levels. A 2012 weight loss study in obese men found that losing 17 pounds increased testosterone levels by 15%. These results were similar to those from a study in European Journal of Endocrinology which showed that reducing weight increases testosterone levels.

A study in the Journal of Endocrinology concluded that low testosterone and diabetes are connected.

Lack of sleep was found to be a major cause of reduction in testosterone levels based on research published in the Journal of American Medical Association. Participants also experienced a decreased sense of wellbeing. The amount of sleep required by an individual varies, but on an average adults need 7 to 9 hours of sleep per night.

Human growth hormone (HGH) is released mainly during sleep and therefore getting a good night's sleep can also increase HGH. As described before, HGH can act synergistically with testosterone to improve overall health. Reducing alcohol intake reduces estrogen and aids in weight loss. Increase water intake, keeping the body hydrated has an overall beneficial effect and also helps with weight loss.

Stress can also cause a reduction in testosterone and HGH levels. Stress reduction techniques such as yoga, meditation, deep breathing and positive visualization should be employed.

Supplements

A search for natural alternatives to testosterone yields a lot of hits on the internet. It should be noted that supplements are not regulated by the FDA and one should always exercise caution when using supplements. There is a risk of hyperpotent or hypopotent dosing. Many supplements may interact with prescription medications. If supplementation is undertaken, ensure that Good Manufacturing Practices (GMP) are used and a good quality control procedure is used to determine the labeled quantity is actually present. Look for endorsements by organizations such as the United States Pharmacopeia (USP).

Dehydroepiandrosterone (DHEA) and Androstenedione

There is a class of supplements that inhibit the enzyme aromatase. Aromatase converts testosterone to estrogen. The most significant one is dehydroepiandrosterone. It is a hormone secreted by the adrenal gland and in the body it is converted to testosterone in men and estrogen in women. Another one is androstenedione. This had gained notoriety in the late 1980s and early 1990s when many athletes were taking it to improve their sports performance.

Tongkat Ali

Tongkat ali is a medicinal plant found in Indonesia and Malaysia. It is also sometimes referred to as Malaysian ginseng. The scientific name of this plant is Eurycoma longifolia. The roots of this plant contain small peptides called eurypeptides and are extracted by cold or hot water extraction. Animal studies have shown that tongkat ali extracts boost testosterone production in rats and other laboratory animals. The mechanism of action appears to be increasing the release rate of testosterone that is bound to sex hormone binding globulin. In a May 2012 study in the journal Andrologia, 76 men with late onset hypogonadism were given 200 mg of tongkat ali extract for 1 month. After 1 month, approximately 72% of the subjects showed normal testosterone levels. Tongkat ali is supposed to reduce cortisol and increase testosterone. A February 2013 study reported in the Journal of the International Society of Sports Nutrition, subjects who were also given 200 mg per day of tongkat ali for 4 weeks showed 16% reduction in cortisol and 37% increase in testosterone levels.

Just as with other supplements, a note of caution. There have been reports of tongkat ali extracts that are fake. Finding a reliable supplement supplier is key.

D-Aspartic Acid

D-aspartic acid is an amino acid that regulates testosterone synthesis. However, its effects are only temporary.

L-Arginine

L-arginine is a non-essential amino acid. But it is very effective in reducing oxidative stress and increasing immunity. In addition, it is a vasodilator, which means it relaxes blood vessels. It can improve libido.

Fenugreek

Fenugreek is an herb and the seeds are used in Indian cooking. But it is a natural testosterone booster. It also increases lean muscle, fat loss and libido. Fenugreek supplements are available, but there are gastrointestinal side effects.

Tribulus Terrstris

Tribulus Terrestris is an herbal extract that is supposed to increase testosterone production.

Pine Pollen

Pine pollen contains plant sterols which are steroid like substances. There are 3 main components that help. Brassinosteroids are powerful growth stimulants. A study conducted in rats showed that brassinosteroids boost testosterone production. Gibberelins are compounds that are structurally similar to testosterone and bind to androgen receptors to increase androgen production. Glutathione transferases are involved in the synthesis of progesterone and testosterone. It has been determined that 10 milligram of pine pollen contains microgram quantities of testosterone, DHEA, androsterone and androstenedione.

Mucuna Pruriens

Mucuna pruriens is a bean from a climbing plant. It is a natural source of L-dopa. L-dopa is metabolized to dopamine. Studies have shown that L-dopa and dopamine stimulate the hypothalamus to release gonadotropin releasing hormone (GRH) which them stimulates the pituitary to release follicle stimulating hormone (FSH) and luteininzing hormone (LH; see Figure 4) ultimately resulting in the Leydig cells secreting testosterone. One animal study and three human studies have showed the effect of Mucuna pruriens on testosterone levels. Mucuna pruriens comes in a powder form.

Vitamin D

Vitamin D is not a vitamin, but a hormone and therefore it has effects on many different systems in the body. Vitamin D is also supposed to help with erectile dysfunction, but its role in testosterone regulation is not known. This is because mixed results were obtained with studies conducted to determine the relationship between Vitamin D (specifically D3) and testosterone levels. A German study saw modest increases in testosterone when Vitamin D3 was administered. However, a Dutch study did not observe any differences in testosterone levels between group given a placebo and group given Vitamin D. There are Vitamin D receptors in the testes and so it is possible that Vitamin D plays a role in testosterone production. Vitamin D3 also reduces effects of metabolic syndrome which can indirectly increase testosterone levels as well. An easier way of getting Vitamin D3 is exposing yourself to the sun. It is synthesized in our body when our skin is exposed to the sun. Studies indicate that Vitamin D synthesis takes half as much time as time required to cause sunburn. So exposing your skin to the sun especially early in the morning can help naturally increase your Vitamin D levels. A study by Boston State Hospital found that when the chest and back were exposed to UV light, testosterone increased 120%, but when the genitals were exposed, testosterone increased 200%.

Vitamin D also helps deliver calcium and phosphorus to the bones and remineralize them.

More light also lowers the levels of melatonin in our body and signals the pituitary to secrete LH and FSH. As discussed earlier, LH and FSH play an important role in testosterone formation.

Zinc

Exercise decreases thyroid hormones and testosterone in sedentary old men. However, taking a zinc supplement increases free and total testosterone. Zinc is naturally present in meats and fish, but also raw milk, raw cheese, beans and yogurt.

Saw Palmetto

This species of palm tree is found in southeastern US. Its fruit is the supplement. It helps raise testosterone levels by inhibiting the conversion of testosterone to dihydrotestosterone. It is also supposed to help with enlarged prostate.

Melatonin

Melatonin is the hormone that regulates sleep. Melatonin also decreases with age. It is a very potent hormone and is effective in quantities less than a milligram, however some might need more. Usually it is recommended to start with the smallest dose and gradually increase it until the right dose is found. It indirectly helps with testosterone levels by helping you sleep better.

There are other supplements that might treat symptoms such as erectile dysfunction, however, scientists are not convinced that supplements really work. Many of these supplements are manufactured overseas and the authenticity of the manufacturer cannot be satisfactorily determined. Another factor to consider is that some supplements may interact with other prescription and over the counter medicines that the person maybe taking, especially older men. So supplementation should be done under medical supervision.

SUMMARY AND CONCLUSIONS

Testosterone is a steroid hormone responsible for the development of sexual characteristics. Since it is a hormone, it affects more than one system in the body. Some of the systems affected by testosterone include the endocrine system, the reproductive system and the cardiovascular system. Effects of testosterone are seen as early as 7 weeks after conception. After birth, testosterone levels gradually increase until they peak at age 20. After age 30, testosterone levels naturally decrease. But in certain individuals, reduction in testosterone levels is caused by a medical condition called hypogonadism. According to the FDA hypogonadism is the only condition for which testosterone replacement therapy (TRT) is approved. However, doctors prescribe TRT for "off label" use. Many doctors do not agree on whether TRT is justified. The primary reason seems to be the lack of large, long term controlled studies. Testosterone is not the fountain of youth. This myth is propagated by fly by night Low T centers. Direct to consumer advertising and qualitative tests influence older men to seek treatment. In 2012, $100 million were spent on advertising for testosterone replacement products and the gross sales that year were $2.2 billion. The projected sales for 2018 are $3.8 billion. A study in the Journal of American Medical Association (JAMA) Internal Medicine found that the number of prescriptions for testosterone from 2001 to 2011 in men 40 and older more than tripled. If a blood test does confirm that testosterone levels are low and additional signs and symptoms such as decrease in muscle mass and strength, decrease in libido and increase in fat deposit around the midsection, TRT may be justified. But it should be undertaken under medical supervision. If treatment is undertaken, there are different dosage forms available. Oral delivery of testosterone is associated with liver toxicity and therefore not advisable. Non-oral routes of administration such as injectable and topical delivery options are available. TRT has been shown to benefit hypogonadal men, but there are severe side effects associated with testosterone therapy. Side effects are seen in reproductive, cardiovascular and hormonal systems. Also friends and family, especially women and children, of patients on TRT are at risk of developing symptoms in case they come in contact with the prescribed drugs or in contact with skin of patients on TRT. Long-term safety studies in large populations have not been carried out. There are non-testosterone drugs such as human chorionic gonadotropin that have also been used. TRT is contraindicated in people with certain pre-existing conditions such as prostate cancer and benign prostatic hyperplasia. Before beginning TRT, a baseline determination of several medical parameters such as blood levels of free and total testosterone, Prostate Specific Antigen, hematocrit, lipid profile, liver function test must be taken. Patients on TRT should also be monitored regularly to make sure that the condition is improving and there is no increase in negative markers of health. TRT works well in a certain segment of people, but it should not be blindly extended to other people who do not fit the criteria for TRT. Supplements can be used to treat low testosterone, but it should be noted that the supplements are not regulated by the FDA. When evaluating a supplement manufacturer, it should be ensured that they use good manufacturing practices and good quality control. Endorsements from organizations such as the

United States Pharmacopeia is a plus. As a general rule, many ailments can be halted or reversed with simple changes in diet, exercise and lifestyle. These approaches are safe and should be tried before other approaches.

ACKNOWLEDGEMENTS

A work of such quality can seldom be achieved by an individual. I would like to acknowledge the help I received from Raymond Aaron Group and my personal book architect, Naval Kumar.

NOTES

NOTES

NOTES